MathStart®
PERIMETER

Racing Around

by Stuart J. Murphy • illustrated by Mike Reed

HarperCollinsPublishers

LEVEL
2

To Gary Facente—who always has lots
of good ideas racing around in his head

—S.J.M.

To Jane, Alex, and Joe

—M.R.

The publisher would like to thank teachers
Patricia Chase, Phyllis Goldman, and Patrick
Hopfensperger for their help in making the
math in MathStart just right for kids.

HarperCollins®, ✚®, and MathStart® are registered trademarks of HarperCollins
Publishers. For more information about the MathStart series, write to HarperCollins
Children's Books, 10 East 53rd Street, New York, NY 10022, or visit our website at
www.mathstartbooks.com.

Bugs incorporated in the MathStart series design were painted by Jon Buller.

Racing Around
Library of Congress Cataloging-in-Publication Data
Murphy, Stuart J.
    Racing around / by Stuart J. Murphy / illustrated by Mike Reed.
        p.   cm.—(MathStart)
    ISBN 0-06-028912-0 — ISBN 0-06-028913-9 (lib. bdg.) — ISBN 0-06-446244-7
(pbk.)
    1. Distances—Measurement—Juvenile literature.   [1. Distances—Measurement.]
I. Reed, Mike, ill.   II. Title.   III. Series.
QC102.M87   2002
530.8—dc21   00-056722

Typography by Elynn Cohen
13 SCP 20 19 18 17 16 15 14 13 12 11
❖ First Edition

Racing Around

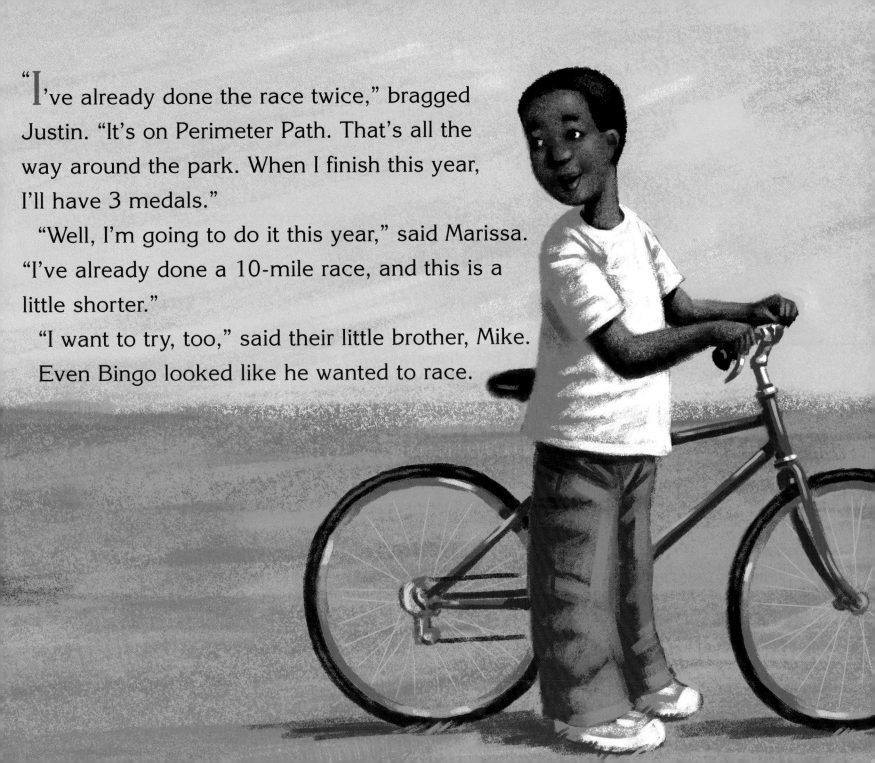

"I've already done the race twice," bragged Justin. "It's on Perimeter Path. That's all the way around the park. When I finish this year, I'll have 3 medals."

"Well, I'm going to do it this year," said Marissa. "I've already done a 10-mile race, and this is a little shorter."

"I want to try, too," said their little brother, Mike. Even Bingo looked like he wanted to race.

"You'll never make it," said Justin with a laugh.

"You'd better wait a few years," Marissa added.

"No way," Mike said. "I can do it. Yesterday I rode my bike all the way around the athletic field. I bet that's almost as long as the race."

"Oh yeah?" said Justin. "Remember that odometer I got for my birthday? It measures distance exactly. I'll check and see how far you went."

Justin started at the corner of the athletic field and rode all the way down 1 side. The odometer read 1 kilometer. Then he rode down the next side. The odometer read 3. That meant the second side was 2 kilometers long. The third side was 1 kilometer again. The odometer was at 4.

And the final side was 2 kilometers.

"I went all the way around, and the perimeter of the field is only 6 kilometers," Justin said.

"That's not even half of 15," said Marissa.

But Mike didn't give up. The next day, when Justin and Marissa were out, he got his dad to sign the permission form.

"I'm going to ride in the race," he announced when his brother and sister came home. "Today I rode all the way around the zoo. I bet that's pretty close to 15 kilometers."

"No way," said Marissa.

"I'll check it tomorrow," said Justin.

ZOO

ENTER

FINISH

START

1

1

2

2

3

12

The next morning Justin rode his bike to the zoo. He set his odometer to 0. When he got to the monkey house, it read 2.

Then it was 1 kilometer to the seal pool. The odometer read 3.

The lion habitat was 1 kilometer away. Now he'd gone 4 kilometers in all.

After 2 more kilometers, Justin was at the birdcage. That made 6. Then 1 more long side and he was back at the entrance.

Marissa, Mike, and Bingo were waiting.

"How far was it?" Mike shouted.

"Just 9 kilometers," Justin answered.

"That's still a long way from 15," Marissa said.

14

Finally, the day of the race arrived. Justin, Marissa, and Mike were at the starting line.

"You wait right here for us, Mike," said Marissa.

But while Justin and Marissa went to get in line, Mike hurried back to the sign-up table. He gave the lady his permission form.

"This is a long race," she said. "Are you sure you can make it all the way around?"

"Yes," said Mike. "I'm sure."

15

Mike stayed at the back of the line. Justin and Marissa never saw him. "Wait right here for me, Bingo," said Mike.

16

"Racers, on your marks!" shouted a man with a megaphone. "Get set! Go!"

All the bikers took off.

4 km

Mike followed the crowd along the first stretch. "This is easy!" he yelled.

Then he started up a small hill. He pumped hard and soon he was at the top.

Mike coasted down the hill. At the bottom was a sign. "Okay!" shouted Mike. "I can do this!"

But when he turned a corner, there was a long, steep hill. He started up . . . and up . . . and up.

I'm never going to get to the top! he thought.

There were only a few other racers in sight. Mike was the last.

20

Finally Mike made it up the hill. He had to stop and rest for a few minutes. But he didn't give up.

The road lay straight in front of him. Mike started pedaling as hard as he could. Ahead, he saw a sign that read 12 kilometers.

"Almost there!" Mike panted.

But then his bike hit a rock. Mike bumped and bounced—and fell.

Mike was tired. He was hot. He was sweaty. And he'd banged his knee. He didn't want to race anymore.

They were right, he thought. *I can't make it all the way around.*

Just then Mike heard something in the distance. It sounded like a dog barking.

It was Bingo!

Bingo raced up and gave Mike a great big lick.
Mike laughed.

"Bingo, did you come to find me?"
he asked. "I guess I'd better finish
the race after all!"

Mike got back on his bike.

4.0 — START

15 km

FINISH

3.0

12 km

1.0

Justin and Marissa had already made it back to the finish line. "Now where's Mike?" Marissa asked. "He was supposed to wait for us here."

"And where's Bingo?" Justin added.

Then they heard the man with the megaphone make an announcement. "Let's hear it for our last rider, folks! He's just coming around the bend of Perimeter Path!"

29

30

Justin and Marissa ran back to the finish line just in time to see Bingo come panting across it. And right beside him was Mike.

"Way to go, little brother!" Justin yelled.

Mike stopped his bike and hopped off.

"I told you!" he said. "I knew I could make it all the way around!"

31

In *Racing Around*, the math concept is perimeter, or the distance around a shape. Perimeter is a geometric concept that helps children understand properties of shapes and distance.

If you would like to have more fun with the math concepts presented in *Racing Around*, here are a few suggestions:

- Prior to reading the story discuss how long a kilometer is (about .6 of a mile). Ten kilometers would be a little more than 6 miles.

- Read the story with your child. As you are reading, have your child trace with his or her finger the perimeter around the athletic field, the zoo, and Perimeter Path. Have the child find the perimeter by adding the lengths of the sides.

- Using a ruler, help the child find the distance around familiar objects in the home, such as picture frames, tabletops, or TV screens. Make a drawing of each object and write the length of each side on the drawing. Then calculate the perimeter.

- Arrange 6 square pieces of paper or tiles so that each square touches at least 1 other along a full side. Count each side that is not touching the side of another square to find the perimeter of your shape. Make some more 6-square shapes and find their perimeters.

perimeter = 14 sides   perimeter = 12 sides

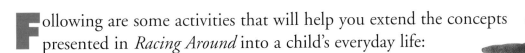

Following are some activities that will help you extend the concepts presented in *Racing Around* into a child's everyday life:

Frame a Picture: Use one of the child's favorite pictures and measure its perimeter. Using construction paper, make a frame for the picture and then measure the perimeter of the frame.

Bike Ride: Plan a bike ride around a favorite site with the child, making sure the ride ends back where it began. Draw a map of the route. Label each section of the route with its length (in blocks, kilometers, or miles) and figure out the perimeter of your route. Enjoy the ride!

Website: Visit the website of a zoo or amusement park. Find a map of the park or zoo and from the map help the child estimate the perimeter.

The following stories include concepts similar to those that are presented in *Racing Around*:

- SPAGHETTI AND MEATBALLS FOR ALL! by Marilyn Burns

- MAPPING PENNY'S WORLD by Loreen Leedy

- MEASURING UP! by Sandra Markle